THROUGH THE EYES & EARS OF THE POET

BY

JAMES E. PEPPERS

ISBN: 978-1-957837-11-6

Contents

About the Author ... 6
Life's Blessing ... 7
Aids ... 8
Happy Birthday Big Mom ... 9
Single Men ... 10
Game Women Play ... 11
Love My Wife Always ... 12
Pray of life ... 13
Summer Sky ... 14
Love So Sweet ... 15
Dreams ... 16
Life Is Good ... 17
Locked Up Behind The Bars ... 18
Lite In One Big Room ... 19
What Do Women Want ... 20
Can We Cuddle Up Tonight? ... 21
Two Hearts Beaten As One ... 22
Life & Love to Carry On ... 23
Someone To Love ... 24
When You Are Down & Out ... 25
Broken Hearts Can Bleed ... 26
My Angel Always ... 27
Pay With Me ... 28
Race ... 29
To The Woman That I Love ... 30

Jail 31

What Is Love? 32

About the Author

I was born in Chicago, Illinois.

I grew up in a family of 16, and I'm the middle child, more like the black sheep of the family, and it's alright. I was raised in the Chicago projects better known as Robert Taylor Housing. I went to Mary C. Terrel School and wasn't too popular with the women until I was about 17.

I have been married twice but failed. Don't know if I would give getting married a third chance, but we will see. I also went to a school called Sturgis E.V.G center, where I learned how to work and make things with wood and plexiglass. Now, I can proudly say that I'm an independent contractor and house cleaner.

I went to school to learn about road construction, and now, I'm certified in that as well. So, this is just a little about me. Rest, you will learn from my writings.

Life's Blessing

God made earth that's the land and open seas and the people like you and me, to help them in that times of need. So, listen my brothers to what I'm saying, lift poor hands and pray to him. And say bye-bye to the serpent. A serpent called Satan. Yes, Satan is a snake, a rat, and a scoundrel, someone who brings you down; but if you pray to the Lord above that serpent, he will not come around. So, lift your hands and pray to Him, for He's always here. Amen.

Written By
James Peppers

Aids

Here's a story of the life so true, mean of all colors,
not me, maybe, you have a good wife at home with a baby or two,
Still do not know what life is all about because
They don't go out looking for a hoe or a prostitute,

Take life as one big chance not to know who is holding hoe's
hand,
Something you can't see, but you will feel in pants.
Guess what? Just find out HIV has got you, man.

Here's a story of a life that is oh oh,
So true when you cheat on your loved one,
Aids is going to get you.
So, go home to your loved ones and leave others in the street
Because life without a loved one is like a gay man or a lady on the
street.

Written By
James Peppers

Happy Birthday Big Mom

There's a lady in my life that I love so much.
She is not my friend, and she is not my wife.
She is a woman that has been there through thick and thin
From the beginning of my life to the end.

There is a lady in my life that I love so very much.
She is sweet and beautiful, a person you can love.
She is loving and caring and so much more.
She is a devoted Christian for all the world to see.

The lady in my life that I love so very much.
She is my grandmother.
That's who she is and today is hers.
As we know, she needed 92 and lived to be more.
Happy birthday, Big Mom

Written By
James Peppers

Single Men

Single men something I know
So very, very well, you lovely and Sad,
Unhappy and sometimes mad,
Wishing I made a woman in my life,
To make me feel glad,

Single men living their life
All alone, a Holiday, a birthday,
And New Year have come ad gone,
and days like that is not like to go on s

Single men whose souls are empty,
Not to love or care for all that's gone,
Someone love, kiss, and hold went,
We come home and open the door.

Written by,
James Peppers

Game Women Play

When looking for men with big muscles
And things some look for nice cars, diamond
Rings some look for men with Forton and fame
And they say men play too many games
Well, some of them do, and that I can say
But not all men are like that but do they really care

Women look for men who curse them out,
And push them around, beats them knocking on the ground
What are the games they love to be around?
Same as the man love to love her down

Woman loves her man, have all different reasons
But most of them will hold the love they need every day,
At least not with the games
Each one of the plays and that think we me the same,
but instead of thinking out who’s real and not playing games

Written by:
James Peppers

Love My Wife Always

My love, my dear, my sweetness, to the woman I always dream about every night I go to sleep and awaken with a smile on my face because I have my wife who loves me so much each and every day.

So here's to you, my love. I love you always, my love, my dear, my sweetheart, my heart for years to come; I'll hold you in my arms, so all the world to see that when I look into your eyes and say yes, you will always be the girl for me.

Written by:
James Peppers

Pray of life

Dear load above to you I pray for I giving the love I share reach and every day And to keep my mind from running to the weakness that heads my way but to keep me strong enough to see all, the pain and misery so thank you Lord for that and the power you give to make and to other people, you people just like me, so I pray to you my lord to guide me all way for the life you gave me each and every day
Amen.

Written By
James Peppers

Summer Sky

The sky is a pretty light blues.
The trees are full of love so.
All the birds hide in an oath the summer breeze,
The grass is nice and soft to lay in.
And daydream into nightfall,
And then stars come out to shine,
Shines down on you while you slowly clear your mind.
And the good life God has given you. So just leave the past behind,
The sky is a peachy light blue,
and the trees are full of leaves to give us.
Wonderful shade when the sun brightens the day,
And send us sometimes very hot temperature to move us along the way,
And make us drink ice-cold lemonade in the hot summer sky.

Written By
James Peppers

Love So Sweet

Sugar is sweet; candy is too. There is a gorgeous Lady that I would like to meet and that my dear is you, the way you walk and the way you talk when you call my name, want you to enter the room and ask me how was my day.

Honey, the thick is a golden brown, and you are the kind of lady I love to be around every day and all through the night, holding you in my arms as we sit by the fire saying the love that we perfectly desire.

Spicey is good if you need that in your life, And red rose, black teddy, massage lotion if that's the case, little body oil that's good to taste me. All I need is you with this passion, a loving face, silky black hair and glamorous smile and body to match, and eyes that shine like the stars at night, making everything just right.

Written By
James Peppers

Dreams

Here I sit looking up in the sky wishing upon the first star that falls, but it doesn't, so I just sit dreaming of a young lady wishing she was here in my arms while laying under the moonlight, listening to the sounds of the night, but it's all a dream, a dream that one day it will come true because all day I think of you.

Dreams are what everybody does; some are sleep and others are wide awake, some dreams are full of heartbreak and pains, some are full of joy and fame and all the money you'll gain, you know life fun as I must say down your dream each night and day.

Here I sit looking up in the sky wondering why women of today, some look for money in a man to them along the way, and some dream of a man they like to be with on a nice cool day to keep them warm and never go away.

Written By
James Peppers

Life Is Good

Life is good, although it seems, a mother, a father, a child in need of the sky, the water, the air we breathe the stars that shine all through the night while staying in bed watching the orange moonlight, wishing you had someone to hold you tight.

Life is good although it seems, the lovely green grass, the birds that sing when the morning comes, A newborn baby who's as sweet as can be a brand-new life, a girl with curls and sandy brown hair, yes, a brand-new baby with a bundle of joy.

Life is good, although it seems love is all I ask for, yes indeed someone to care for a person like me. Life is good, yes, we all know, God says come on we have to go for he gave us a life we all must share, cause when God calls you home, yes, we all will be there, yes life is good, and I do care.

Written By
James Peppers

Locked Up Behind The Bars

Bricks falling on your head witch no one sees, no unce but your cally in cell block G, talking through a vent, that sit by the right nigh on the well saying to each other, mom he said, I think I've been here since the middle of the fall, hey what did you do to get in hare man!
I was speeding, and I hit the deer, but the dear won't hurt, I didn't hit him that hard, I knock him in that dirt he got up and ran so I did too, all you young brothers that go for a ride click, click here we go in cell block number five.

Written By
James Peppers

Lite In One Big Room

We all are men in one big room, some are waiting to take a shower, and some are pushing a broom, others are laid back waiting to be groomed that's a haircut or a shark because judgment day is coming for some not too soon, but for others, they spending there likes inside this and big room.

Written By
James Peppers

What Do Women Want

What do women want? Well, let's see, for one, it's not me a sweet and Loving gentleman, that's what I'll be, well I've tried that, and it didn't work.

What do women want, man will never know; it's a wedding ring, a dozen roses, or a car to get around and so they can just get up and go?

What do women want? Can you tell me, please? It is the man who would beat them down to their knees, or the kind of man who treats them like diet and doesn't care if you are hurt.

What do women want, or do they know? Is it another woman, Or a sugar daddy that's always on the go! OR a man who puts you at the street corner and says go make me some money, well I'm not a pimp or player, I'll just be me, the man with a prayer.

Written By
James Peppers

Can We Cuddle Up Tonight?

Roses are so many colors, and the ones I like the most are red and yellow, white is okay but so very plain say I like a little fun in my life wouldn't? You ride the same! But when it's time to cuddle with you, I'm a man who's serious and like no games.

Violets are blue, and the sea is too; that's the way I'll be if I can't be with you; your skin is so soft and smooth, and a cute little smile that brightens the day! So I'll like to be your lover when your man is gone away.

Your hair is silky smooth, and beautiful brown eyes, and Lips so sweet, and a heart of gold maybe one day I can kiss you from your head to your wed, you know where it goes, and hold you in my arms nice and tight, as we cuddle up each and every night.

Written By
James Peppers

Two Hearts Beaten As One

Roses are red, and some flowers are blue, my heart is just thinking of you, sitting over there so far but yet so near looking at your loving face and your sandy brown hair soft smooth skin, I lust for your body all the again kissing you so gently white holding you so close. Loving every inch of your body, that's what I like most watching you tingle with expectations; as we submit to each other, our hearts will beat as one.

Written By
James Peppers

Life & Love to Carry On

Heavenly another high above, the father we all love and pray for the sins we have done, and to ask for a blessing for others who are doing wrong. We all know that God gave us life, soul, and the air we breathe to carry. Until it's time for us to come home to the man, to the person who loves us more than anything in this world, and now someone we love and adore so deeply, is going home to the person that gave us life and love to carry on, so sleep on a sweet lady and take your rest. We love you, but God loves you best.

Written By
James Peppers

Someone To Love

I am a man whose life is very bored, because there is no one to
kiss, to hug or to love anymore, a woman so soft and sweet,
someone to love that's what I like to meet.

Someone to love indeed, someone to love that's what I need,
A woman who loves me for me, the man within
Not for what he spends.

A woman full of love and drunk, I agreed, honest and caring.
As a relationship goes devoted to each other,
Everyone can see,
Someone to love a brother like me.

Written By
James Peppers

When You Are Down & Out

When you are down and out and nowhere to go, who do you turn to your brother, father,
Hey friend, no, not even your bro because none of them will show you which way to go,

So, who do you turn to when you are down and out? Your mother, that's who she will show you the way out because nine of ten answers will be put in God's hands, for he is the only one who knows.

When you are down and out, he's the only way to go because he gave you love and life so he will know,
When you are down and out, he will comfort the soul.

Written By
James Peppers

Broken Hearts Can Bleed

Hearts are broken in so many different ways by the way you talk, and what you say that makes a person's happiness and into a sad and lonely heartbroken way full of anger, to everybody that gets in the way any time of the day.

Hearts are broken in so many different ways, by the wife that's cheating in with his best friend at the Wedstar Inn is where her husband and his lover first began, so its life goes on a broken heart will bleed throughout until they can become as one.

Hearts are broken in so many different ways by the lover or the best friend, someone you can always love through thick and thin, a soulmate for life, but who am I to say, I am just a person with a broken heart that bleeds as the days are long and the nights are very short.

Written By
James Peppers

My Angel Always

There is a woman I love so much, she the apple of my eye, and the sweet smell of jasmine every time she walks by and tells me, honey I love you sometimes being a tear to my eye that makes me say thank you Lord for sending you sending her my way.

There is a woman I Love each and every day, yes, I'll truly say my friend, my Lover, my girl, my wife, I will love her and love her always, and when I am down, she puts her arms around me and tells me baby don't worry, I'll be with you always.

Written By
James Peppers

Pay With Me

People kneel down and pray every day, some full of light, some a little gray; But the Lord will hear no matter what you say for he will save you to the will of his pray so give him a chance and kneel down with me so he or show you the way, the way of life that's his way because He is the Lord, Lord he will have you the right way, so kneel down with me to say the prayer, each and every day.

Written By
James Peppers

Race

People of a different race, a man, a woman, or whatever the case may be, find themselves in places where they love to be, but If you go to the state of the town, you will see that they don't like Italians, Mexicans, and Blacks like me! I am a person who loves people no matter what race they are, so to all of you who like to get around traveling from city to city and town to town, well here I stand to let you know, Town or state you don't want to go because some of the people will let you know, there's only one race and others have got to go.

I'm sorry to say people of the town are that way, so when my job is done in 15 days, I will be so happy to get away…

Written By
James Peppers

To The Woman That I Love

This is to the woman that I love,
So very much as the day goes on,
I wish so much to touch, to kiss, and to hold you into my arms,
To care for and live as the days go on.

This is to the woman that I love so much as the day comes to an end,
And the night creeps in,
And the smile on your face as we snuggle up once again.

This is to the woman that I love so much,
As her heart and mine beat as one, as I look into her eyes
While we just sit there, here's a rose to hold until I get
There to show my love, that I always care.

Written By
James Peppers

Jail

Jail is a place where some people are locked behind bars,
A steel door that's where they be,
Closed down from the world,
Yes, sir, and now that's me

A man with a warm heart something they cannot see,
I do what I can to help a person in need,
Now I am locked up like a dog waiting to be set here,
Yes, jail is a place some people can see,
Begging for their life, for one day we will be set free,
For a breath of clean air, a mother, a father, and yes, a person like me.

Written By
James Peppers

What Is Love?

Love, what is love?
Well, love is not a character,
And love is not a poem,
But love is what you give,
To every child that's born.

Love is what you give, and
Sometimes what you receive,
But if we use it right,
Love will never leave.

Love comes out of the heart, and
Love comes out of the soul; if
You treasure the love you have,
Your love will never grow old.

Love is in the air, and
Love is all around
Love is what God gives
To every walk on,
His earthly ground.

Written By
James Peppers

Made in the USA
Monee, IL
04 November 2026

67376092R00020